梦幻的答案之书

小欣 著

北京燕山出版社

图书在版编目（CIP）数据

梦幻的答案之书 / 小欣著. -- 北京：北京燕山出版社，2022.10
ISBN 978-7-5402-6644-8

Ⅰ．①梦… Ⅱ．①小… Ⅲ．①人生哲学－通俗读物 Ⅳ．① B821-49

中国版本图书馆 CIP 数据核字（2022）第 169114 号

梦幻的答案之书

作　　者：小　欣
出 品 人：余　言
责任编辑：李　涛
特约编辑：赵　迎
封面设计：陈秋含
出版发行：北京燕山出版社有限公司
地　　址：北京市丰台区东铁匠营芜子坑 138 号 C 座
邮政编码：100079
电　　话：（010）65240430
印　　刷：长沙鸿发印务实业有限公司
开　　本：889mm×1194mm　1/32
印　　张：13
字　　数：80 千字
版　　次：2022 年 10 月第 1 版
印　　次：2022 年 10 月第 1 次印刷
书　　号：ISBN 978-7-5402-6644-8
定　　价：45.00 元

◪ 使用指南 ◪

· · ·

1. 放下心中杂念，保持内心平静，拿起这本合着的书，放在你的腿上或桌子上。

2. 默想你心中最想问的问题，尽量是简单的封闭式问题。

   例如："这个工作适合我吗？"

   "我该追求她吗？"

   "他真的适合我吗？"

3. 深呼吸，翻开书，睁眼看这一页的答案。

4. 希望对你的人生有所帮助。

☆ ·::::· 你所有的问题，这本书都能为你解答 ·::::· ☆

有风险，但也有机会

In the shadow of risks there are also
opportunities

Answers

十拿九稳

Bet on it

Answers

尽早行动

Do it early

千万别孤注一掷

Don't put all your eggs in one basket

Answers

你的行动会使事情变得更好

Your actions will improve things

Answers

那可不一定

It's not necessarily the case

Answers

为什么不呢

Why not

Answers

难以预料

There is no telling

Answers

最好再等等

Better to wait

Answers

结果可能会令人吃惊

The result may be startling

Answers

请教一下这方面的专家

Consult an expert in this field

Answers

你需要适应

You should adapt yourself to it

Answers

它会带来好运

It will bring good luck

Answers

享受这个过程

Enjoy the process

Answers

要有耐心

Be patient

Answers

这会影响其他人对你的看法

It will affect how others see you

Answers

尽力而为

Do your best

Answers

你会很高兴你这么做了

You'll be very happy to do so

Answers

此时不宜

It is not the time to do it at the moment

Answers

勿忘初心

Stay true to your original mind

Answers

只要你按照被告知的方法去做

If you do as you're told

Answers

能做好就做，否则干脆别做

Do it well, or forget it

Answers

不要要求太多

Don't ask too much

Answers

不要半途而废

Never do things by halves

Answers

谨慎行事

Be cautious

Answers

大声说出来

Say it out loudly

Answers

不要犹豫

Do not hesitate

Answers

无法保证

There is no guarantee

Answers

相信自己，你可以的

Believe in yourself, you can manage it

Answers

情况很快就会有所改变

Things will change soon

Answers

不要被情绪左右

Do not get caught up in your emotions

Answers

寻求更多的选择

Seek out more options

Answers

仅仅是时机不对

It's just not the right time

Answers

先不做决定

Deal with it later

Answers

值得冒险一试

It's worth the risk

Answers

很麻烦

Very troublesome

Answers

再坚持一下

Just hold on

Answers

全力以赴

Go all out

Answers

这不是你可以控制的

It's beyond your control

Answers

你需要主动争取

You should be aggressive

Answers

你必须妥协

You will have to give in

Answers

再考虑一下吧

Think about it

Answers

有好运

Have a good luck

Answers

对其他人保密

Keep secret from others

Answers

相信你最初的想法

Trust your original thought

Answers

最好专注于你的工作

Mind your won business

Answers

大胆一点

Be bold

Answers

先完成其他的事情

Finish something else first

Answers

你可能遭到反对

You may have opposition

Answers

需要付出巨大的努力

It requires great efforts

Answers

让自己先休息一下

Take a break first

Answers

机会难得

It's a rare chance

Answers

不妥

It's not proper

Answers

等待更好的机会

It will be good for you

Answers

它会让你受益

It will sustain you

Answers

实际点

Be practical

Answers

结果将是好的

The outcome will be positive

Answers

不要担心

Don't worry

Answers

你一定能获得支持

You are sure to have support

Answers

坚持一定会有回报

Persistence will pay off

Answers

不会令你失望

I will not let you down

Answers

不要迫于压力而仓促行事

Do not be act in haste under pressure

Answers

不值一搏

It is not worth a Struggle

Answers

遵从内心

Stay true to your own mind

Answers

极有可能发生糟糕的事情

We are in for it

Answers

是的

Yes

Answers

答案就在你内心深处

The answer is in your mind

Answers

佛系点儿

Let it go

Answers

你并非真的在意

You do not really care

一年之后这将不再重要

It will not matter any more in one

year

Answers

不要浪费你的时间

Don't waste your time

Answers

可能会超乎寻常地好

It can be exceptionally good

Answers

数到 10，再问一次

Count to 10, and ask again

Answers

一定非常成功

It must be a great success

Answers

请保持冷静，这样才能做出最好的
选择

To stay calm and make the
best decision

Answers

你会后悔的

You will regret it

Answers

梦里什么都有

You can have everything in a dream

Answers

当然

Of course

Answers

相信你的直觉

Trust your intuition

Answers

把它当作一个机会

Take it as an opportunity

Answers

问问长辈的建议

Ask your elders for advice

Answers

不

NO

Answers

你的行动会推动事情的发展

Your actions will get things better

Answers

别犯傻

Don't be silly

Answers

本页前的第八个答案

The Eighth answer before this page

Answers

事情正朝着好的方向发展

Things are turning for better

Answers

玩得开心就好

Just have fun

不需要

No, not necessary

Answers

保持头脑清醒

Keep a cool head

Answers

不确定因素有点多

There are quite a few uncertainties

Answers

去做其他的事情

Go to do something else

Answers

把重心转移到开心的事情上

Focus on something else that can make you happy

Answers

马上就会有解决办法的

There will be a solution soon

Answers

毋庸置疑

There is no doubt

Answers

决定了就去做

Do as you decide

Answers

删除记忆

Delet the memory

Answers

问问自己什么是最重要的

Ask yourself what is most important

Answers

摒弃首选方案

Give up the preferred alternatives

Answers

试着卖个萌

Try to be cute

维持现状

Maintain the status quo

Answers

你将为之付出代价

You will pay for it

Answers

对意外提前做好准备

Prepare for the unexpected

Answers

它没有什么意义

It makes no sense

Answers

保持开放的心态

Keep an open mind

Answers

将会有一个阻碍需要克服

There will be an obstacle to overcome

Answers

慢慢来，不要急于求成

Take your time

Answers

不要忽略细节

Don't ignore the details

Answers

它不值得努力

It's not worth the effort

Answers

木已成舟

It is a done deal

Answers

不能失败

Can't fail

Answers

结果将会是积极的

The result will be positive

Answers

机不可失，时不再来

Now or never

Answers

马上行动

Act now

Answers

记得分享

Remember to share

Answers

这事儿不靠谱

It is not reliable

Answers

与众不同

Be different

Answers

现在已经很好了

It's good now

Answers

是的，但别勉强

Yes, but don't force it

Answers

你觉得呢

What do you think

Answers

不妨先去度个假

Maybe take a vacation first

Answers

多点耐心

More patience

Answers

信任

Trust

Answers

制定一个新的计划

Make a new plan

Answers

你有 B 计划吗

Do you have Plan B

Answers

这会很美好

It can be great

Answers

需要找到更多解决办法

Find more solutions

Answers

事情开始变得有趣了

Thing are getting interesting

Answers

对别人大方点

Be generous to others

Answers

很快会发生令人期待的事情哦

Something desirable is about
to happen

Answers

从来没有

Never

Answers

抱一抱自己

Embrace yourself

Answers

还没到绝境

It's not the end yet

Answers

无论你做什么，事情都会继续下去

Whatever you do, it's gonna

go on

Answers

关注你的日常生活

Pay attention to your daily life

Answers

该做打算了

It's time to make plans

Answers

就这么干吧

Just do it

Answers

只能做一次

Only once

Answers

欣然接受它

Embrace it

Answers

笑着面对

Face it with a smile

Answers

继续前进

To move forward

做点让你开心的事情吧

Do what makes you happy

Answers

默数十秒再问我一次吧

Ask me again in tens of seconds

Answers

一笑而过

Laugh it off

Answers

去改变

To change

Answers

不要害怕

Don't be afraid

Answers

想法太多，选择太少

More ideas than decisions

Answers

这是显而易见的

It's obvious

Answers

需要多花点时间

It's going to take a little bit more time

Answers

保存你的实力

Save your strength

学会妥协

Learn to compromise

Answers

改变不了世界，就改变你自己

You can't change the world, but you
can change yourself

Answers

别傻傻地等待

Don't wait around

Answers

不要陷得太深

Don't get too deep

Answers

最佳方案不一定可行

The best solution is not necessarily feasible

Answers

观望

On the sideline

Answers

平常心对待

Take things as they are

Answers

似乎没问题

It seems all right

Answers

那就算了吧

Let's forget it

Answers

不明智

Not wise

Answers

还有另一种可能

There's another possibility

Answers

列明原因

List reasons

Answers

着眼未来

Focus on the future

Answers

记录下来

Write it down

Answers

别赌

Don't bet on

Answers

荒谬

The absurd

Answers

别不自量力

Don't overreach yourself

Answers

错的

In the wrong

Answers

你需要找个人来帮你

You need someone to help you

Answers

没有

There is no

Answers

可行

Practicable

Answers

醒醒吧，别做梦了

Wake up, Stop dreaming

Answers

看看事物的另一面

Look at the opposite side of things

Answers

不要忽略身边的人

Don't ignore the people around you

Answers

不要单独行动

Don't do it by yourself

Answers

去倾听

To listen to

Answers

不要顾虑太多

Don't worry too much

Answers

等待一下会有更好的选择

Wait a moment and you'll have a

better choice

Answers

随波逐流未必是件好事

Going with the flow is not necessarily

a good thing

Answers

培养一个新的爱好吧

Develop a new hobby

Answers

把重心放在工作 / 学习上

Focus on your work/study

Answers

选择容易走的路

Take the easy road

Answers

问天问地问别人，不如问问你自己

Don't ask others, ask yourself

Answers

接受一些改变

Accept some changes

Answers

有贵人助你成功

Someone will help you succeed

Answers

守口如瓶

Keep your mouth shut

Answers

采纳智者的建议

Take the suggestions of the wise

Answers

情况不明了

The situation is not clear

Answers

独善其身

Do your own things well

Answers

主动出击，人生迥异

Take the initiative, and make a

difference in your life

Answers

休息，休息一会儿

Take a break

Answers

不妨赌一把

Why not take a bet

Answers

协作

Cooperation

Answers

阻止

Stop it

Answers

灵活应对

Deal with flexibility

Answers

照别人的话去做

Obey others' opinion

Answers

看开一点

Take it easy

Answers

你要了解真相

You need to know the truth

Answers

看看再说

Let's see

Answers

可能会有连锁反应

There could be a chain reaction

Answers

以后再处理

Handle it later

Answers

不值得斗争

It's not worth fighting

Answers

明天再来试试

Try again tomorrow

Answers

享受这次体验

Enjoy this experience

Answers

人心不足蛇吞象

No man is content with his lot

Answers

小心为妙

Be careful

Answers

注意身后

Watch your back

Answers